PREDATORS AND PREY

Alyxx Meléndez

Consultants

Darrin Lunde
Collection Manager
National Museum of Natural History

Cheryl Lane, M.Ed.
Seventh Grade Science Teacher
Chino Valley Unified School District

Michelle Wertman, M.S.Ed.
Literacy Specialist
New York City Public Schools

Publishing Credits

Rachelle Cracchiolo, M.S.Ed., *Publisher*
Emily R. Smith, M.A.Ed., *SVP of Content Development*
Véronique Bos, *VP of Creative*
Dani Neiley, *Editor*
Robin Erickson, *Senior Art Director*

Smithsonian Enterprises

Avery Naughton, *Licensing Coordinator*
Paige Towler, *Editorial Lead*
Jill Corcoran, *Senior Director, Licensed Publishing*
Brigid Ferraro, *Vice President of New Business and Licensing*
Carol LeBlanc, *President*

Image Credits: p.13 Alamy Stock Photo; p.24 NOAA NMFS SWFSC; p.25 (top) Marc Dando; p.25 Dave Mellinger/Oregon State University; all other images from iStock and/or Shutterstock, or in the public domain.

Library of Congress Cataloging-in-Publication Data

Names: Melendez, Alyxx, author. | Smithsonian Institution, contributor.
Title: Predators and prey / Alyxx Meléndez.
Description: Huntington Beach, CA : TCM, Teacher Created Materials, Inc., [2025] | Includes index. | Audience: Ages 10+ | Summary: ""In every ecosystem around the world, predators hunt prey for food. Predators and prey have evolved to have unique features and strategies for survival. Explore how predators and prey on land and in water try to find food-and avoid becoming food themselves""-- Provided by publisher.
Identifiers: LCCN 2024018897 (print) | LCCN 2024018898 (ebook) | ISBN 9798765968628 (paperback) | ISBN 9798765968703 (ebook)
Subjects: LCSH: Predation (Biology)--Juvenile literature.
Classification: LCC QL758 .M45 2025 (print) | LCC QL758 (ebook) | DDC 591.5/3--dc23/eng/20240517
LC record available at https://lccn.loc.gov/2024018897
LC ebook record available at https://lccn.loc.gov/2024018898

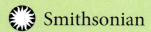

5482 Argosy Avenue
Huntington Beach, CA 92649
www.tcmpub.com
ISBN 979-8-7659-6862-8

© 2025 Teacher Created Materials, Inc.
Printed by: 51497
Printed in : China

© 2025 Smithsonian Institution. The name "Smithsonian" and the Smithsonian logo are registered trademarks owned by the Smithsonian Institution.

This book may not be reproduced or distributed in any way without prior written consent from the publisher.

Table of Contents

The Hunter and the Hunted. 4

Delicate Balancing Act 6

Predator and Prey Behavior 10

Special Features 16

Underwater Pursuits. 22

The Cycle of Survival 26

STEAM Challenge 28

Glossary. 30

Index. 31

Career Advice 32

The Hunter and the Hunted

A lion slinks through the tall grasses in an African savanna. Today, the lion is hunting a distant warthog. The warthog is quick, but it is no match for the lion's speed. The lion sprints toward the warthog and pounces. In an instant, the lion sinks its claws into the warthog's thick, hairy hide and delivers a final bite to the warthog's neck.

Predators, like this lion, are animals that hunt and eat other animals. The animals they pursue, like this warthog, are their prey. Predators and prey exist all around the world. They come in many shapes, colors, and sizes. They all have different abilities, features, and behaviors that help them survive.

Lions typically stalk their prey from a covered position before chasing them down.

In every ecosystem, there is a balance of predators and prey. Ecosystems are made up of the living and nonliving parts of an environment. Living things include **organisms**, such as plants, animals, and fungi. Nonliving things include sunlight, soil, and water sources. These parts are all connected. Diagrams called *food chains* show how predators and prey get the energy they need to survive.

Predators and prey are important in each ecosystem around the world. Exploring how predators and prey live and behave in different environments provides insight into how most animals function.

What do spiders, frogs, and grizzly bears have in common? They are all predators!

Delicate Balancing Act

Ecosystems are built on a complex web of relationships. Producers, consumers, and decomposers make up the living things in an ecosystem. Producers are plants, and they make their own food. Consumers eat plants, animals, or both. Predators and prey fall into this group. Decomposers break down dead animals and plants and return their **nutrients** to the soil. These three groups function together in a delicate balance.

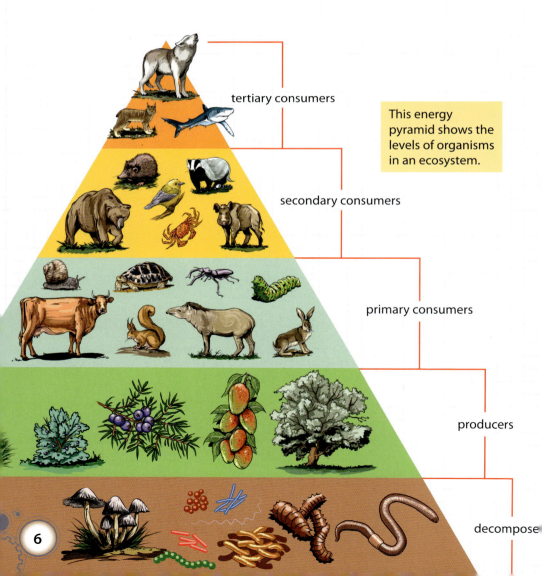

This energy pyramid shows the levels of organisms in an ecosystem.

Caught in the Food Chain

A food chain shows how living things depend on one another for food. Arrows show the connections between predators and their prey. All predators and prey are consumers. There are three different types of consumers.

First, there are primary consumers. These animals only eat plant matter, and they are called *herbivores*. Elk, rabbits, and voles are primary consumers. They are all prey.

Secondary consumers are next. These animals can be herbivores or omnivores. They might eat only plants, or both plants and animals. Certain birds, raccoons, and foxes are secondary consumers. These animals can be both predators and prey.

Tertiary consumers are the last type of consumer. These animals are always at the highest level of a food chain. In most cases, they are carnivores who only eat meat. These animals are all predators.

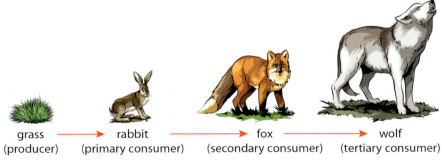

grass ⟶ rabbit ⟶ fox ⟶ wolf
(producer)　(primary consumer)　(secondary consumer)　(tertiary consumer)

FUN FACT

When a consumer has killed an animal and eaten their fill, other consumers may feast on the **carcass**. These consumers are called *scavengers*. They pick the bones clean before decomposers come along. Vultures are well-known scavengers.

Breaking the Balance

Both the living and nonliving parts of an ecosystem are **interdependent**. When something changes, it affects all other parts of the ecosystem—sometimes in negative ways. These effects then ripple through a food chain.

A prime example of this occurred when gray wolves disappeared from Yellowstone National Park. By 1926, the park had zero gray wolves due to overhunting. No one expected the consequences that followed.

First, the park's elk population flourished. This might sound like a good thing, but it caused an imbalance. While grizzly bears and coyotes also ate elk, gray wolves were their main predators. The wolves had been keeping the elk population in check. Next, elk began to eat all the willow, aspen, and cottonwood trees. Elk ate the trees faster than they could grow. For many years, these trees had trouble growing beyond saplings. As a consequence, once the elk dominated, beavers were left with hardly any saplings. This was an important food source for them, and their population suffered.

During hunting, a wolf pack spreads out to surround its prey.

Over time, scientists realized they had to bring back gray wolves. So, in 1995, they reintroduced gray wolves into the park. The population started small, and it grew to more than 100 over the next two decades. Now, a more natural balance among animals and plant life has been restored.

On average, elk eat 9 kilograms (20 pounds) of plant material a day!

Bringing Back the Bison

Like gray wolves, bison were hunted in Yellowstone National Park. The National Park Service estimates that there were roughly two dozen bison left in 1902. With lots of hard work over many decades, their numbers increased again. As of 2022, park staff counted around 5,900 bison.

Predator and Prey Behavior

Predators and prey, whether large or small, behave in different ways. Their instincts are to survive, and every action they take affects their existence.

Apex Predators

Every food chain has an **apex** predator. Apex predators are so dominant that other animals tend to stay clear of them. As a result, they can hunt without being hunted themselves. These predators can be found around the world. On land, big cats, bears, and wolves are some of the most successful apex predators. In **aquatic** ecosystems, sharks and orcas reign supreme.

Great white sharks have sharp, jagged teeth that help them consume their prey.

eagle talons

Predators have evolved to have different **adaptations** for hunting. On land, most apex predators have sharp, thick claws. Their teeth are sharp enough to pierce and tear flesh. In the air, **birds of prey** dive and swoop during their hunts. They have hook-shaped beaks and thick claws called *talons*. Some aquatic predators may lack claws, but they use their teeth instead.

Sometimes, when apex predators go out to hunt, they have to leave their **young** behind. Their young may lack the size and skill for adequate self-defense. For example, jaguar cubs must watch out for anacondas. Lion cubs are under constant threat from hyenas. Because of these dangers, not all young apex predators make it to adulthood. But when they do, no other species can stop them.

lion cubs

hyena

Small Hunters

Not all predators are large like grizzly bears. Insects and **arachnids** may be small, but these tiny predators have evolved to have special hunting techniques. They attack their prey in different ways.

Some insects, such as ants, form armies to invade other insects' burrows. Ants will go to war with termites, other types of ants, or even ants of the same species. Certain types of ants work together to take down larger animals, such as scorpions and crabs. They use their scissor-like jaws to cut into the joints of their prey.

Ants attack scorpions by biting or stinging them.

Bull ants use large mandibles to crush or cut their food.

Some spiders use their webs to hunt in innovative ways. Net-casting spiders hold sticky nets in their four front legs. Then, they reach out and grab their prey. Diving bell spiders create tightly wound webs underwater. Bubbles of oxygen stay in their webs, giving these spiders a safe place to consume their prey.

An Australian net-casting spider spins a web.

Other spiders act like ninjas, attacking prey by jumping on them. Some jumping spiders use their silk-like bungee cords to anchor themselves. Their powerful legs work like springs as they jump. The silk allows them to fling themselves with precision at unsuspecting bugs. Before portia jumping spiders take a leap, they like to play a trick. They fool other, larger spiders by strumming the larger spiders' webs with their legs. This mimics the feeling of a trapped, struggling insect, and the larger spiders come running. Then, portia jumping spiders pounce!

portia jumping spider

ENGINEERING

Spidery Structures

Spiders' webs are amazing feats of architecture. Some engineers have looked at spider webs for inspiration. The Moore Building in Miami, Florida, is one example of this. Sculptures inside the building have been compared to a spider's web. The large, white shapes that link each floor make visitors feel like tiny insects.

13

How Does Prey React?

Because danger could be lurking anywhere, prey need to be skillful at getting away from a threat. These animals act on their instincts to stay safe from predators. Here are just a few examples of how they react to danger.

Prairie dogs live in burrows.

Smaller prey, including many **rodents**, may use their burrows to hide. Birds maneuver through the air to evade capture. In rainforests, monkeys can scamper up trees. Ground-dwellers, such as deer, can run fast to get away. These forms of escape are necessary for survival.

Because large prey animals are easy to spot, they stick together in groups called *herds*. Animals in the middle of the herd are the safest. As the herd moves, their predators follow them, trying to pick off animals around the edges. Predators never hesitate to pounce on young or sick members of a herd. That's because they are always the easiest targets.

Herding behavior is a crucial strategy in grasslands. With very few trees, there are not many places for prey to hide. So, staying in herds and reacting quickly to threats protects the fittest members of a herd. African and Asian elephants behave in this way. So do African buffalo, Asian water buffalo, and North American bison.

FUN FACT

Some prey have unique ways of defending themselves. Porcupines, for instance, have sharp quills. The quills easily detach into a predator's skin. Puffer fish have a similarly spiky method of protection. When these fish sense an attack, they puff up their bodies. Pointy spikes appear, and predators have two choices: get hurt or swim away.

Female African elephants live in herds with their young, while male elephants live on their own or in smaller groups.

15

Special Features

Predators and prey have different adaptations for survival. Their eye shapes, colors, **toxins**, and abilities to blend in help them hunt and stay safe from danger.

Vision

The eyes of predators and prey give them specific advantages. Predators' eyes tend to face forward, just like humans. This positioning helps them better judge the distance to their prey. Meanwhile, prey have eyes on the sides of their heads. They can see almost everywhere around their bodies and detect approaching danger.

Some animals can see more wavelengths of light than others. In the Arctic tundra, caribou and reindeer can see ultraviolet, or UV, light. This type of light is invisible to human eyes. UV light reflects off white surfaces, such as snowy ground. Meanwhile, green **lichens** and dark-colored wolves absorb UV light. So, caribou and reindeer can spot them in an instant.

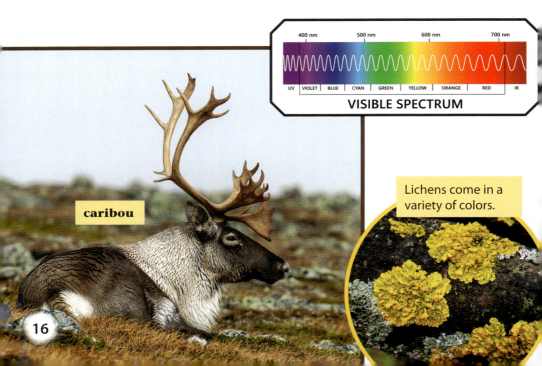

Lichens come in a variety of colors.

On the other hand, some animals can see fewer wavelengths of light than others. While a tiger's bright fur would stick out to a human, it would go unnoticed by a water buffalo. That's because these animals see the world differently—in shades of black, grey, and white. In fact, most **mammals** can't tell the difference between shades of red and green. So, a tiger's orange coat and green plants would appear to look the same, and most mammals would not sense a threat.

what humans see

what most mammals see

ARTS

Undercover Camera Crews

To capture animals on camera, photographers and filmmakers have to think like animals. They might hide in fake leaves or perch in trees to get the perfect shot. At night, they use low-light cameras that make good use of moonlight, just like a predator's eyes would. In total darkness, they use **infrared** cameras.

Poisons and Venoms

Animals' fur, skin, and feather colors can communicate information. For example, flashy feathers on male birds draw attention from female birds. But bright colors on some animals may warn others to stay away.

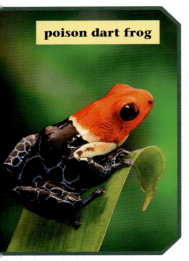
poison dart frog

Some **poisonous** prey have vibrant colors to ward off predators. For example, if predators attack colorful and shiny poison dart frogs, they will end up with mouthfuls of poison. There are more than 100 species of this tiny **amphibian**. Each one has a unique pattern that helps them attract mates—and warn predators.

Unlike frogs, poisonous birds don't usually make their poison. Instead, they take it from poisonous bugs. Blue-capped ifrits are one example. They put poison from the insects they eat into their feathers to protect themselves. Their blue, shiny heads serve as a warning to predators to stay away.

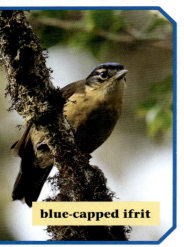
blue-capped ifrit

Many predators are **venomous**. This means that they produce toxins, usually through fangs or stingers. Then, they inject it into their prey. Some toxins kill prey instantly. Others disorient or paralyze prey. Colorful venomous predators can be found in a range of environments. The yellow-and-black stripes of bees, wasps, hornets, and yellow jackets warn other animals of their toxic stings. In the ocean, blue-ringed octopuses have yellow skin with rings that change colors when they sense threats. If their prey gets too close, the octopus bites and injects toxic saliva into the bite mark.

Blue-ringed octopus venom is highly toxic and causes paralysis.

Immune to Pain

When bark scorpions sting, their venom causes extreme pain to animals and humans. But the tiny grasshopper mouse is one exception. When they are bitten, they don't feel any pain. These mice evolved to be resistant to the venom. This adaptation allows them to eat the scorpions.

SCIENCE

Mimicry

People can be copycats—and so can prey. Certain animals have evolved to look like others, which gives them the benefit of protection from predators. This resemblance is called *mimicry*.

Viceroy butterflies know all about mimicry. With their orange and black wings, they look similar to monarch butterflies. Birds know to stay clear of monarchs because they are poisonous. And birds avoid viceroy butterflies, too, even though they aren't poisonous. That's because the two species look so similar. It's only when you look closely that you notice the differences.

Some animals can mimic entirely different types of animals. For example, a few types of caterpillars have large black spots on both sides of their heads. Their snakelike appearance frightens birds.

viceroy butterfly

monarch butterfly

Camouflage

Animals can use **camouflage** to blend in with their surroundings. They may look like their environment or parts of it, such as plants, rocks, or soil. This way, predators can't see where they are. Many kinds of insects have this ability. Praying mantises blend in with leaves or twigs. And the orchid mantis blends in with beautiful orchid flowers.

dead leaf mantis

orchid mantis

Cuttlefish and some octopuses take camouflage a step further. Depending on their environment, they can change the color of their skin. They have special **iridescent** skin cells. These cells reflect light, and the light changes colors when it is viewed from different angles. To change color, these animals can stretch or shrink their skin cells.

cuttlefish

Underwater Pursuits

In aquatic environments, predators and prey have similar adaptations compared to animals on land. Seals have sensitive whiskers to detect motion, just like cats. These whiskers help seals locate swirls of water where fish have recently swam. And both snakes and sharks have special ways of opening their mouths. Sharks can detach their upper jaws so they can take large bites. This resembles the way certain snakes dislocate their lower jaws to swallow their prey whole.

In water, just like on land, predators have different ways of attacking prey. Swordfish and many sharks swim directly through groups of fish, snapping them up as they go. Sea lions, much like lions on land, surround their prey and force them to move in a tightly packed group. Then, they hunt the weakest animals along the edges.

To stay safe from threats, fish swim together in groups. This behavior is similar to how mammals form herds. Although fish along the edges will be eaten by predators, most of them stay safe in the center of the group. Loose, disorganized clusters of fish are called *shoals*. Some species can form highly coordinated groups called *schools*. Every fish in a school moves in the same direction at the same time to form complicated shapes. They do this to protect themselves from predators.

California sea lions hunt a school of fish.

Humpback whales hunt herring.

FUN FACT

Humpback whales are huge and slow compared to other whales. To catch their prey, they rely on their blowholes. They blow bubbles out of their blowholes, churning up the water and disorienting fish. Then, they take as many fish as they can into their mouths.

Marine Mammal Attacks

While sharks hunt alone or in pairs, orcas and dolphins form larger groups called *pods*. They use teamwork to hunt their prey. Each pod has its own hunting strategy.

Some bottlenose dolphins off the coast of South Carolina developed a unique hunting method. During low tide, they surround fish and create a powerful wave with their bodies. This wave pushes them toward the muddy shore. Then, they fling themselves onto the shore with their mouths open. They grab all the fish they can and slide back into the water.

In the Arctic, orcas hunt seals. Seals sit on floating ice sheets for rest, shelter, and mating. But they end up defenseless when orcas attack. Orcas rush at the ice sheets, which creates a big wave. Then, they dive below the water, pushing the wave forward with their tails. The wave goes over the ice and crashes onto the seals, knocking them into the water.

Weddell seal and orca

Orcas have also developed a method to hunt great white sharks. First, they ram into the sharks' bodies to stun them. Next, they swim underneath the sharks and flip them upside down. This disorients the sharks so much that they stop moving. At this point, orcas can bite into the sharks. This process can happen quickly, sometimes in a matter of minutes.

Orcas can hunt prey in groups or on their own.

TECHNOLOGY

Hydrophones

Scientists use hydrophones to listen to and record sounds in the ocean. These underwater microphones can detect sounds made by animals. This is how scientists know that orcas use **echolocation** to hunt. Orcas make clicking noises underwater to locate their prey.

25

The Cycle of Survival

Every animal in the world shares the same goal: to find food and avoid becoming food. Predators have evolved to have the most effective ways to hunt. Prey have evolved to use methods that keep themselves safe from danger. Whether these animals live underwater, on land, or in the air, they act on their instincts to survive.

All predators and prey have evolved to have unique features for survival. This is true whether they are big or small. Flashy colors on their fur or feathers are a way of communicating with other animals. Their vision gives them special advantages. Some prey are poisonous, providing a built-in defense mechanism. Some predators use venom to attack their prey. These adaptations affect how predators and prey react in different situations.

In the end, all predators and prey become food themselves in death. Their energy goes to decomposers. With these tiny life forms around, no morsel of energy goes to waste. Decomposers return composers' nutrients to the soil. This allows new plants to grow, which then feed consumers—who often become prey. And a select few of them become predators. The endless cycle between predators and prey goes on and on.

In all environments and ecosystems around the world, predators hunt prey in different ways.

STEAM CHALLENGE

Define the Problem

Spiders, like engineers, design and build complex structures called *webs* to trap and collect their prey. Can you think like an arachnid to create an amazing and effective web to help maintain balance in an ecosystem? Your task is to design and build a model web that is able to trap and hold three different plastic insects within the web.

Constraints: You may only use the materials provided to you. Your web must be at least 25 centimeters (10 inches) across.

Criteria: The web must be able to be held upright. It must be able to trap and hold three different insects.

Research and Brainstorm
What is the relationship between predators and prey? How do spiders make webs? How do they use webs to trap their prey?

Design and Build
Sketch two or more designs for your spider web. Label the parts and materials. Choose the design you think will work best. Then, build your spider web.

Test and Improve
Share your spider web with others. Point out any special features on your web. Explain why you chose these particular materials to build the web. Demonstrate how your web works by using it to trap three different insects. Was your design successful? How do you know? How can you modify your design to trap smaller insects or more insects at once? Modify your design and rebuild as needed. Reassess how well it meets the criteria.

Reflect and Share
What makes your web design unique? How can you use ideas from other designs to improve your team's design? What did you learn that you can apply to other challenges?

Glossary

adaptations—changes that make organisms better at surviving in their environments

amphibian—a type of cold-blooded animal that spends part of its life in water and part of its life on land

apex—the top or highest point of something

aquatic—growing or living in or often found in water

arachnids—a group of invertebrate animals, such as spiders and scorpions

birds of prey—predatory, carnivorous birds that have sharp talons

camouflage—coloration that blends with an environment or surroundings

carcass—a dead body, usually of an animal

echolocation—the process of locating objects through reflected sound

infrared—a type of light that is invisible to human eyes

interdependent—dependent upon one another

iridescent—shimmery colors that change when seen from different angles

lichens—plantlike organisms growing on rocks or trees that are a food source for animals

mammals—warm-blooded animals, usually with hair or fur, who feed their young with milk

nutrients—substances that organisms eat for survival

organisms—living things such as plants, animals, fungi, and bacteria

poisonous—producing a toxic substance that can cause illness or death

rodents—mammals with strong, constantly growing front teeth

toxins—substances produced by living organisms that can cause illness or death

venomous—having or containing venom, which can be injected into others through fangs or stingers

young—animal babies

Index

African buffalo, 14
animal vision, 16–17, 26
apex predators, 10–11
arachnids, 12
Arctic, 16, 24
Asian water buffalo, 14
camouflage, 21, 31
carnivores, 7
consumers, 6–7, 26
decomposers, 6–7, 26
food chain, 5, 7–8, 10
grasslands, 14
herbivores, 7

hydrophones, 25
insects, 12–13, 18, 21
marine mammals, 24–25
Miami, 13
mimicry, 20
Moore Building, 13
omnivores, 7
poison, 18, 20, 26
producers, 6–7
scavengers, 7
South Carolina, 24
venom, 18–19, 26
Yellowstone National Park, 8–9

A pallid scops owl uses camouflage to blend with tree bark.

CAREER ADVICE
from Smithsonian

Do you dream of working with animals?

Here are some tips to keep in mind for the future.

"Volunteer and intern with animals. Many places (such as zoos and aquariums) look for people who have hands-on experience handling and caring for animals. Some of my colleagues started their animal experience by volunteering at local animal shelters in their area."

– *Bayley McKeon*, Ocean Education Specialist, Smithsonian National Museum of Natural History

"If you want to study animals, translate your curiosity into adventurousness and be open to diving in wherever you see professionals already at work. Aquariums, zoos, sanctuaries, and conservation groups in your area are all great places to build the connections that can help you grow a career."

– *Alia Payne*, Ocean Education Specialist, Smithsonian National Museum of Natural History